THE POETRY OF CARBON

The Poetry of Carbon

Walter the Educator™

SKB

Silent King Books a WhichHead Imprint

Copyright © 2023 by Walter the Educator

All rights reserved. No part of this book may be reproduced in any manner whatsoever without written permission except in the case of brief quotations embodied in critical articles and reviews.

First Printing, 2023

Disclaimer
This book is a literary work; poems are not about specific persons, locations, situations, and/or circumstances. This book is for entertainment and informational purposes only. The author and publisher offer this information without warranties expressed or implied. No matter the grounds, neither the author nor the publisher will be accountable for any losses, injuries, or other damages caused by the reader's use of this book. The use of this book acknowledges an understanding and acceptance of this disclaimer.

*dedicated to all the chemistry lovers,
like myself, across the world*

CONTENTS

Dedication v

Why I Created This Book? 1

One - Ingenious Art 2

Two - Symphony Of Atoms 4

Three - Story Of Connection 6

Four - CO2 8

Five - Reigns Supreme 9

Six - True Superstar 11

Seven - Grand Display 13

Eight - Diamonds To Coal 15

Nine - Web Of Life 17

Ten - Unsurpassed 19

Eleven - Connecting You And Me 21

Twelve - Carbon's Grace 23

Thirteen - Power Of Carbon	. . . .	25
Fourteen - Living Space	. . . .	27
Fifteen - Every Living Cell	. . . .	29
Sixteen - Grand Affair	. . . .	31
Seventeen - Crystal Clear	. . . .	33
Eighteen - Carbon's Legacy	. . . .	35
Nineteen - For All To See	. . . .	37
Twenty - Essence Of You And Me	. . . . . .	39
Twenty-One - Foundation Of Existence	. . . .	41
Twenty-Two - My Friend	. . . .	43
Twenty-Three - Depths Of Earth	. . . . . .	45
Twenty-Four - Building Blocks	. . . .	47
Twenty-Five - Many Ways	. . . .	49
Twenty-Six - Secret Of Life	. . . .	50
Twenty-Seven - Learn And To Grow	. . . . .	52
Twenty-Eight - Dazzling Dream	. . . . . .	54
Twenty-Nine - Without You	. . . .	56
Thirty - Star Of Chemistry's Cosmic Embrace	. . . .	58
Thirty-One - Versatility	. . . .	60

Thirty-Two - Eternal Allure 62

Thirty-Three - Defies Limitation 64

Thirty-Four - Weaves Its Magic 66

Thirty-Five - Ever-present 68

Thirty-Six - Backbone 70

About The Author 72

WHY I CREATED THIS BOOK?

Creating a poetry book about the chemistry element of Carbon offers a unique opportunity to explore the beauty and complexity of this fundamental element in a creative and imaginative way. Carbon is the building block of life, present in all organic compounds, and plays a crucial role in our existence. By employing poetry, I can delve into its various forms, properties, and interactions, capturing the essence of Carbon's versatility and significance. Through metaphor and symbolism, I can evoke emotions, highlight the interconnectedness of life, and celebrate the wonders of science. This book can bridge the gap between art and science, appealing to both poetry enthusiasts and chemistry lovers.

ONE

INGENIOUS ART

In the realm of chemistry's embrace,
A stellar element takes its place,
Carbon, the foundation of life's design,
A cosmic bond that will forever shine.

Six protons dance in its atomic core,
Neutrons and electrons, an intricate lore,
A carbon atom, so elegantly arranged,
A symphony of forces, beautifully exchanged.

Carbon, the architect of organic birth,
In every living form, its presence unearth,
From the towering trees to the creatures that roam,
Carbon's versatility, a masterpiece to own.

Diamonds, precious jewels, born of its might,
Carbon's compact structure, a mesmerizing sight,
Yet in the depths of coal, its humble guise,
Carbon's transformation, a wondrous surprise.

In the depths of time, carbon's story is told,
From ancient fossils to diamonds, rare and bold,
In the carbon cycle, a dance of give and take,
Creating equilibrium, life's delicate stake.

From the whispers of photosynthesis' song,
To the strength of steel, so steadfast and strong,
Carbon's bonds, unyielding and true,
Connecting the world, through and through.

So let us marvel at carbon's beauty profound,
Its essence in every creature around,
A testament to nature's ingenious art,
Carbon, the element that sets life apart.

TWO

SYMPHONY OF ATOMS

In the realm of elements, Carbon reigns supreme,
A foundation of life, a dreamer's cherished theme.
In every living being, its essence holds sway,
From the tiniest cell to the grandest display.

Carbon, oh Carbon, versatile and true,
A chameleon of atoms, changing its hue.
A diamond's shimmer, a charcoal's embrace,
In every form, it leaves its trace.

From the depths of the earth, it rises with might,
In coal and in oil, burning bright.
Unveiling its secrets, through the carbon cycle's dance,
Connecting the world, in a cosmic romance.

Through photosynthesis, nature's divine art,
It weaves a tapestry, connecting every part.
From the leaves of green forests to the depths of

the sea,
Carbon whispers its tale, setting life free.

It binds with hydrogen, oxygen, and more,
Creating the molecules we've come to adore.
In sugars and fats, its energy unfurled,
Carbon's touch, the essence of our world.

In the depths of the ocean, it makes its abode,
Supporting the ecosystems, life's intricate code.
From coral reefs to the creatures that dwell,
Carbon's presence, a story to tell.

Oh Carbon, the beauty you possess,
In every molecule, you caress.
From the flutter of butterfly wings to the soaring of birds,
Carbon's magic, in every word.

So let us marvel at this element sublime,
The building block of life, throughout space and time.
For in Carbon's embrace, we find our start,
A symphony of atoms, setting life apart.

THREE

STORY OF CONNECTION

In the realm of atoms, a chameleon stands,
Carbon, the weaver of life's intricate strands.
A versatile element, so wondrous and grand,
Foundation of existence, across every land.

In diamonds, it dazzles, with brilliance untamed,
A crystal's allure, forever unchained.
In coal's dark embrace, a story untold,
A treasure of ages, hidden in the cold.

Carbon, the architect of organic design,
Building molecules, a symphony so fine.
In sugars and fats, it dances with grace,
Fueling life's fire, in the human race.

From forests to oceans, it connects every part,
Carbon's hand, weaving life's tapestry of art.

In the breath of trees, it whispers a song,
A symphony of green, where life may belong.
    In the depths of the ocean, where wonders lie deep,
Carbon sustains life, a secret it keeps.
From coral reefs to plankton's embrace,
Carbon's embrace, a delicate space.
    So let us celebrate this element divine,
Carbon, the essence that makes life align.
In every molecule, it plays its role,
A story of connection, from atoms to soul.

FOUR

CO_2

Carbon, oh Carbon, a foundation of life,
Your versatile nature, cuts like a knife,
From diamonds to graphite, you take on new forms,
Your beauty and strength, like a raging storm.

In the depths of the ocean, you're sequestered away,
Locked in the earth's crust, you patiently stay,
But when life needs you, you're always on call,
A vital connection, for one and for all.

You bind with the oxygen, to form CO_2,
A crucial molecule, for plants to renew,
Photosynthesis, your grand contribution,
A source of life, in symbiosis and fusion.

From the forests to the cities, you keep us alive,
A cycle of life, that will always thrive,
Carbon, oh Carbon, a wonder to behold,
A true testament, to nature's own mold.

FIVE

REIGNS SUPREME

In the realm of elements, Carbon reigns supreme,
A versatile soul, a poet's dream.
With six protons, six neutrons, and electrons to sway,
It dances through nature in a mesmerizing display.

Carbon, the foundation of life's grand design,
Connecting all beings, intertwining like a vine.
From the depths of the ocean to the sky above,
It weaves its magic, an eternal love.

In the heart of a diamond, it sparkles and gleams,
A crystalline beauty, beyond all our dreams.
In the coal's embrace, it holds warmth and might,
Fueling the fires that illuminate the night.

Carbon, the architect of organic creation,
Building molecules with boundless fascination.
In the living cells, its chains unfold,
Forming the essence of stories yet untold.

From the whispering trees to the buzzing bees,
Carbon breathes life into the gentlest breeze.
In the wings of a butterfly, it takes flight,
Painting the world with colors so bright.

Carbon, the keeper of the carbon cycle's dance,
Absorbing, releasing, in a mesmerizing trance.
Through photosynthesis, it captures the sun's grace,
Creating a symphony of life in every place.

From the tallest mountains to the deepest sea,
Carbon brings harmony, a celestial decree.
It molds the Earth with its gentle touch,
Creating a world that's cherished so much.

So let us honor this element divine,
Carbon, the poet of atoms, so sublime.
For in its elegance, we find a profound truth,
That in unity and diversity, lies the essence of youth.

SIX

TRUE SUPERSTAR

In the realm of chemistry, behold Carbon's might,
A versatile element, radiant and bright.
Building blocks of life, in countless arrays,
Carbon weaves its magic, in mysterious ways.

From diamonds so precious, with brilliance untold,
To pencil lead, scribbling stories of old.
Graphene, a marvel, a lattice so fine,
Conducting electricity, pure and divine.

Carbon, the backbone, of organic design,
Binding atoms together, in a symphony so fine.
In cells and tissues, it dances and weaves,
Creating the beauty, that life so achieves.

In the depths of the ocean, where darkness resides,
Carbon finds solace, in the secrets it hides.
For in the abyss, where life takes its form,
Carbon sustains ecosystems, weathering every storm.

From corals to plankton, and creatures so small,
Carbon fuels their existence, one and all.
In the web of life, where connections are made,
Carbon is the thread, that will never fade.

So let us marvel, at Carbon's profound,
Its versatility, in every compound.
In the world of chemistry, a true superstar,
Carbon, the element, that shines from afar.

SEVEN

GRAND DISPLAY

In the depths of Earth's ancient bosom,
A treasure lies, both bold and awesome.
Carbon, the element of life's grand stage,
Weaving wonders with its atomic wage.

From diamonds, sparkling in celestial light,
To coal, dark and fierce, with fires ignite.
Carbon's allure, a dance of pure delight,
Transforming, transcending, day into night.

In the realm of molecules, it intertwines,
With oxygen, hydrogen, and nitrogen's lines.
Building the chains of life's intricate design,
A symphony of atoms, harmonious and fine.

In the breath we take, in every living cell,
Carbon weaves its magic, a tale to tell.
A backbone strong, a foundation true,
Life's essence captured, in its carbon hue.

From the towering trees, to the humble earthworm,
Carbon's presence weaves a delicate firm.
In bones and tissues, it forms the very core,
Fueling the web of life with its eternal lore.

So let us celebrate this elemental star,
Carbon, the gatekeeper, near and far.
For in its essence, life finds its way,
Forever bound, in Carbon's grand display.

EIGHT

DIAMONDS TO COAL

In the realm of life, Carbon reigns supreme,
A master of transformation, a creator of dreams.
From the depths of ancient Earth it did arise,
Forging the chains of existence, a brilliant surprise.

In every living cell, Carbon finds its place,
Weaving the tapestry of life with grace.
From the towering trees to the creatures that roam,
Carbon's touch is felt, a foundation for home.

In the forests, it dances with the leaves,
Capturing the sunlight, life it conceives.
Through photosynthesis, it breathes in the air,
Nurturing the world with its tender care.

In the oceans, it journeys with the waves,
Building the coral reefs, a haven it craves.
From the smallest plankton to the mighty whale,
Carbon's embrace is felt, an eternal tale.

In the soil, it toils with the roots,
Feeding the plants, bearing the fruits.
From the grains of wheat to the fields of green,
Carbon's essence is seen, a cycle unseen.

From diamonds to coal, it takes many forms,
A chameleon of elements, it defies norms.
With its versatile bonds and infinite chains,
Carbon's legacy remains, forever sustained.

So let us marvel at Carbon's grand design,
A building block of life, so divine.
In every breath we take, in every beat of our hearts,
Carbon's symphony plays, nature's finest art.

NINE

WEB OF LIFE

In the dance of atoms, Carbon takes its place,
A versatile element, with elegance and grace.
Within the carbon cycle, it weaves its intricate thread,
Connecting all of life, from the living to the dead.

In the lush embrace of nature's verdant green,
Carbon breathes life into the world, unseen.
Through photosynthesis, it captures the sun's light,
Transforming it into energy, with all its might.

From the towering trees to the smallest blade of grass,
Carbon builds the foundations of life, en masse.
In the depths of the oceans, where life teems,
Carbon dwells, orchestrating nature's dreams.

In the diamond's gleam and the coal's dark fire,
Carbon's allure never fails to inspire.

From ancient fossils to the pencil in your hand,
Carbon's forms are vast, like shifting sand.
 In the molecules of life, it forms the backbone,
Binding DNA and proteins, never to disown.
In every cell, it plays a vital role,
Sustaining the rhythms of life, heart and soul.
 Carbon, the architect of ecosystems grand,
Connecting all living things, hand in hand.
In the web of life, it weaves its intricate design,
A testament to its beauty, so divine.
 So let us celebrate Carbon, this wondrous element,
For its significance and versatility, so inherent.
In every breath we take and every beat of our heart,
Carbon's presence is a work of art.

TEN

UNSURPASSED

In the realm of atoms, Carbon reigns supreme,
A versatile element, a chemist's dream.
With six protons, a nucleus strong,
Its presence in nature, a lifelong song.

Carbon, the architect of life's grand design,
Building blocks of organisms, both yours and mine.
In the depths of Earth, as coal it resides,
A treasure trove of energy, it provides.

From diamonds to graphite, a stark contrast,
Carbon's forms are varied, unsurpassed.
In the depths of the ocean, it dances with ease,
Creating coral reefs, a vibrant masterpiece.

In the cycle of life, Carbon takes its part,
A web of existence, a delicate art.
In DNA and proteins, it weaves its charm,
A symphony of molecules, a cosmic alarm.

From the tallest tree to the smallest bug,
Carbon connects us all, like a snug hug.
It cycles through ecosystems, in a graceful dance,
Nurturing life with every single chance.

So let us celebrate Carbon, this element divine,
For in its presence, life's wonders align.
From the depths of the Earth to the blue sky above,
Carbon, the foundation of life, we'll forever love.

ELEVEN

CONNECTING YOU AND ME

In the cosmos of elements, a star so bright,
Carbon, the foundation of life's infinite flight.
Four valence electrons, a dance of pure grace,
Binding atoms together, creating a cosmic embrace.

In diamonds, it sparkles, a crystalline delight,
Reflecting the universe with shimmering light.
In the depths of coal mines, its energy hides,
A treasure untold, where darkness subsides.

In the air that we breathe, it silently weaves,
Connecting all living things, like intricate leaves.
In the whispers of forests, it dances and sways,
Nurturing life, in its mystical ways.

Carbon, the alchemist, transforming with ease,
From graphite to diamonds, it shapes as it pleases.

In the DNA helix, it stores our very code,
A blueprint of life, forever bestowed.
   Unseen, yet vital, in cycles it roams,
From atmosphere to oceans, it finds its way home.
From the depths of the Earth, to the heights of the sky,
Carbon, the conductor, orchestrating life's sigh.
   So let us marvel at this element so grand,
For without carbon, life would not stand.
From the tiniest microbe to the tallest tree,
Carbon, the essence, connecting you and me.

TWELVE

CARBON'S GRACE

In every breath, in every tree,
Lies the essence of Carbon, wild and free.
From diamond's gleam to coal's dark hue,
Carbon weaves its tales, both old and new.

In the depths of the Earth, where time stands still,
Carbon lays dormant, patient, until
Volcanoes erupt with fiery might,
Releasing Carbon's essence, pure and bright.

Through the rivers and streams, it finds its way,
Feeding the plants that sway and play.
In leaves of green and petals of flowers,
Carbon breathes life, hour after hour.

From the oceans deep to the skies above,
Carbon dances, a symbol of love.
In creatures big and creatures small,
Carbon weaves its magic, touching us all.

In the beating heart, in the souls we share,
Carbon lingers, a presence rare.
It binds us together, like a gentle embrace,
A reminder of our interconnected space.

So let us marvel at Carbon's grace,
Its power to transform, to leave a trace.
For in this element, we find our worth,
A reminder of our place on this Earth.

THIRTEEN

POWER OF CARBON

In the realm of elements, Carbon shines bright,
A versatile gem, a maker of life's light.
Its atomic dance weaves the fabric of all,
From the depths of the ocean to the trees so tall.

Carbon, the builder, the architect of form,
In molecules and compounds, it becomes the norm.
From diamonds to graphite, it takes many guises,
Changing its structure, revealing its surprises.

In the depths of the earth, it patiently lies,
As coal and as oil, a treasure in disguise.
Through heat and through pressure, it transforms its state,
Emerging as fuel, a source we appreciate.

In living beings, Carbon finds its home,
As the backbone of life, it freely roams.

In every living cell, it plays a vital role,
Connecting the threads that make us whole.

DNA, the code of life, relies on Carbon's touch,
A symphony of molecules, it orchestrates so much.
Proteins, enzymes, and all that we hold dear,
Carbon's fingerprints are everywhere, crystal clear.

From the trees in the forest to the creatures of the sea,
Carbon weaves its magic, connecting you and me.
In the cycles of nature, it dances with grace,
Nurturing life's web, leaving no trace.

So let us celebrate this element supreme,
Carbon, the foundation of life's grand scheme.
With gratitude and awe, let us embrace,
The power of Carbon, our world's saving grace.

FOURTEEN

LIVING SPACE

In the realm of atoms, a jewel doth reside,
Carbon, the element, so versatile and wide.
With four valence electrons, it yearns to bond,
Creating life's tapestry, forever beyond.

From diamonds to coal, in darkness it gleams,
Carbon's allure, a poet's fondest dreams.
In graphite's embrace, it whispers its tale,
Of pencils and sketches, where thoughts set sail.

A backbone of life, DNA's sacred code,
Carbon weaves the story, where secrets unfold.
In molecules complex, it dances with grace,
Bridging the elements, in an intricate embrace.

In the depths of the ocean, where corals reside,
Carbon builds reefs, with colors so wide.
A symphony of life, in vibrant array,
Carbon's transformation, on display.

Compounds it shapes, with a masterful hand,
Forming the building blocks, where life expands.
From sugars to fats, and proteins so grand,
Carbon's chemistry, a symphony unplanned.

In the web of life, it connects us all,
From the tallest tree to the creatures so small.
Carbon, the thread that binds us together,
In the circle of life, it will endure forever.

So let us celebrate, this element so fine,
Carbon, the foundation, where life intertwines.
A testament to nature's infinite grace,
Carbon, the essence, in every living space.

FIFTEEN

EVERY LIVING CELL

In the realm of atoms, let Carbon shine,
A transformative element, so divine.
With six protons, it stands strong and true,
Creating compounds, life it can imbue.

It bonds with itself, forming sturdy chains,
A backbone of molecules, nature's reins.
From diamonds to coal, it takes many forms,
An elemental chameleon that transforms.

In organic matter, its presence is vast,
From the depths of the oceans to the skies so vast.
DNA, the code of life, it does hold,
Carbon's versatility, a story yet untold.

In ecosystems, it weaves a vital thread,
From plants to animals, it provides their bread.
Photosynthesis, a dance so grand,
Carbon capturing sunlight, a symbiotic band.

It cycles through nature, relentless and true,
From the soil to the air, and back anew.
Respiration and decomposition, a cosmic ballet,
Carbon's role in the web of life, a wondrous display.

From the depths of the Earth to the heights of the trees,
Carbon connects it all, like a gentle breeze.
It's the foundation of life, the thread that binds,
Carbon, the element that intertwines.

So let us marvel at Carbon's grand design,
A symbol of connectedness, so sublime.
In every breath we take, in every living cell,
Carbon's presence, a tale we must tell.

SIXTEEN

GRAND AFFAIR

In the realm of atoms, Carbon claims its reign,
A versatile element, nature's eternal chain.
From diamond's brilliance to the humble coal,
Carbon's allure is beyond measure, a story untold.

In the vast expanse of the cosmos it resides,
Forming the building blocks, where life abides.
Complex compounds, woven with delicate art,
Carbon's touch shapes the world, from end to start.

In the depths of the Earth, where darkness lies,
Carbon transforms, a metamorphosis in disguise.
Pressure and time, a symphony of creation,
Carbon's magic gives birth to every formation.

In the realm of the living, Carbon finds its place,
A precious element, connecting every trace.
From the towering trees to the smallest cell,
Carbon's presence weaves the web of life so well.

In the dance of molecules, Carbon takes the lead,
Bonds are formed, a connection we desperately need.
From carbohydrates to the DNA strand,
Carbon's symphony conducts life's grandstand.

Through photosynthesis, Carbon captures the light,
Transforming energy, a wondrous sight.
Green leaves breathe in, exhaling life's breath,
Carbon's cycle, a dance of birth and death.

In every breath we take, Carbon is there,
A symbol of connectedness, beyond compare.
From the depths of the oceans to the heights of the sky,
Carbon unites us all, a truth we can't deny.

Oh, Carbon, foundation of life's grand affair,
In your essence, we find wonder and care.
A symbol of transformation, resilience, and grace,
Carbon, you hold the key to life's eternal embrace.

SEVENTEEN

CRYSTAL CLEAR

In the heart of life's intricate dance,
Stands Carbon, with elegance and grace,
A master builder, an alchemist's dream,
Weaving the tapestry of existence it seems.

From the depths of the Earth's ancient womb,
Carbon emerges, a celestial bloom,
Through fire and pressure, it transforms its state,
Creating diamonds, pure and iridescently great.

Carbon, the element of life's embrace,
Building blocks of nature, in every place,
In the lush green forests, it breathes and thrives,
Sustaining all creatures with each cycle it drives.

From the whispering breeze through the leaves,
To the songs of birds that the morning weaves,
Carbon connects us, in a delicate bond,
A symphony of life, forever beyond.

In the shimmering rivers and oceans wide,
Carbon dances, a constant tide,
From the coral reefs to the mighty whale,
It nurtures, protects, and will never fail.

Carbon, the secret behind every smile,
In the laughter of children, pure and wild,
In the tears we shed, both joy and sorrow,
It binds us together, today and tomorrow.

From the beating hearts of lovers entwined,
To the ancient trees that stand refined,
Carbon weaves its magic, a cosmic art,
Uniting all life, from the end to the start.

Oh, Carbon, the architect of life's grand design,
With your touch, the universe aligns,
In every breath we take, you're ever near,
A testament to your presence, so crystal clear.

So let us celebrate this element divine,
For it's Carbon that makes us, yours and mine,
A symphony of atoms, in harmony we sway,
Forever connected, in Carbon's embrace we'll stay.

EIGHTEEN

CARBON'S LEGACY

In the realm of life's symphony,
Carbon dances with harmony.
A building block of all we see,
A cosmic thread, connecting you and me.
　In the heart of DNA's embrace,
Carbon weaves the threads of life's grace.
A double helix, a code divine,
Carbon's touch, a tapestry so fine.
　Proteins, the architects of form,
Carbon's bonds, a structure to adorn.
From enzymes to hormones, a grand array,
Carbon's versatility, on full display.
　From the depths of the ocean's floor,
To coral reefs, where life's wonders soar,
Carbon nurtures, a vital role to play,
In the web of life, where creatures sway.

Through photosynthesis' gentle art,
Carbon breathes life into every part.
From greenest leaves to towering trees,
Carbon's dance sustains all living seas.

In every breath, Carbon weaves its spell,
A bridge between heaven and earthly shell.
From the whispers of a gentle breeze,
To the roar of waves upon the seas.

From tears of joy to tears of grief,
Carbon shares in every emotion, brief.
Love's passion, Carbon's flame,
Binding souls, in eternal fame.

And in the vastness of the universe's sprawl,
Carbon connects us, one and all.
From stardust born, we find our worth,
For Carbon is the essence of our birth.

So let us cherish this element divine,
For it is Carbon that makes us intertwine.
In this tapestry of life we weave,
Carbon's legacy, forever to believe.

NINETEEN

FOR ALL TO SEE

In nature's grand design, Carbon weaves its spell,
A master of transformation, it weaves the tale so well.
In the depths of Earth's embrace, it slumbers deep,
Till time and heat awaken its secrets to keep.

From humble coal to diamonds rare,
Carbon's essence permeates the air.
In molecules of life, it finds its home,
Building blocks of existence, from ocean to dome.

In photosynthesis, it plays a vital role,
Harnessing the sun's energy, it captures the soul.
Through green leaves, it dances, a dance of light,
Creating sustenance, a symphony so bright.

Carbon, the architect of life's grand scheme,
In every living thing, it reigns supreme.
From the tallest trees to the tiniest seed,
It connects us all, a universal creed.

So let us celebrate this element sublime,
Carbon, the foundation of the passage of time.
Bound by its presence, we stand tall and free,
A testament to its power, for all to see.

TWENTY

ESSENCE OF YOU AND ME

Carbon, oh carbon, a versatile element
Found in many forms, from diamonds to soot
Tetravalent, it bonds with ease
Creating compounds, over ten million, no less
    But carbon's role is greater still
It's the basis of life, a vital part of the deal
Photosynthesis, the process of life
Carbon dioxide, the source of strife
    Yet in every breath we take, carbon's there
Uniting all living things, a bond we share
From the air we breathe to the food we eat
Carbon, oh carbon, so essential and neat
    So let us cherish this element so grand
For without carbon, life wouldn't stand

It's the building block of all we see
Carbon, oh carbon, the essence of you and me.

TWENTY-ONE

FOUNDATION OF EXISTENCE

In the depths of Earth's vast realm,
A secret lies, a timeless helm.
Carbon, the element of grace,
A versatile soul, a cosmic embrace.

From humble coal to sparkling diamond,
Carbon's forms are truly awe-inspiring.
It weaves its magic in every living thing,
A symphony of life, an eternal spring.

In the whispers of the ancient trees,
Carbon dances with the gentle breeze.
Through photosynthesis, it weaves its spell,
Transforming sunlight, a magic so well.

In the bones that bear our mortal frame,
Carbon stands tall, a pillar of the same.

From the depths of the oceans to the sky above,
Carbon unites us, the essence of love.
   It binds us in our very breath,
A constant reminder of our earthly quest.
In every molecule, in every cell,
Carbon's presence, a story to tell.
   Oh, Carbon, you are nature's art,
A masterpiece, a beating heart.
In your elegance, we find solace and might,
The foundation of existence, shining bright.

TWENTY-TWO

MY FRIEND

In the realm of life, Carbon weaves its spell,
A chemical dance, where all things dwell.
From humblest plant to the grandest beast,
Carbon is the essence that binds the feast.

In photosynthesis, it takes its flight,
Harnessing the sun's pure, golden light.
From carbon dioxide, it builds a dream,
Transforming it into life's vibrant stream.

From leaf to stem, from root to flower,
Carbon's touch wields nature's power.
In every breath we take, in every beat,
Carbon's fingerprints, so bittersweet.

In diamonds, it shimmers with radiant grace,
A crystalline testament to time and space.
In coal and oil, it fuels our might,
A testament to its transformative might.

Carbon, oh Carbon, so versatile and true,
The foundation on which life does accrue.
From chemistry's embrace, to nature's decree,
Carbon, my friend, you are life's key.

TWENTY-THREE

DEPTHS OF EARTH

In the dance of life, Carbon takes its place,
A symphony of elements, in perfect embrace.
From the depths of Earth to the sky so blue,
Carbon weaves its magic, creating life anew.

Within its core, a universe resides,
Atoms entwined, like lovers' ties.
Carbon, the builder, of structures grand,
In every living thing, it leaves its hand.

It breathes in the forests, so lush and green,
Photosynthesis, a masterpiece unseen.
With sunlight's touch, it captures the rays,
Transforming energy, in nature's praise.

From the mighty oak to the smallest seed,
Carbon binds them all, fulfilling their need.
In the air we breathe, it whispers its name,
A silent servant, in the cycle of life's game.

But Carbon's tale doesn't end with trees,
For it takes many forms, as life decrees.
In diamonds, it sparkles, a treasure untold,
A symbol of love, more precious than gold.

In graphite's embrace, it writes its story,
A humble servant, in every pencil's glory.
And in the depths of coal, it lies in wait,
A source of power, to shape our fate.

Oh, Carbon, the chameleon, with many a face,
In every form, you leave a trace.
From the depths of Earth to the stars above,
You're a testament to nature's love.

So let us marvel at Carbon's might,
The essence of life, shining so bright.
For in its atoms, a world is born,
A symphony of elements, forever sworn.

TWENTY-FOUR

BUILDING BLOCKS

In the realm of photosynthesis,
Where life's energy begins to rise,
Carbon dances with the sun's embrace,
A symphony of life, in every space.

Through the greenest leaves it flows,
Absorbing light, nature's repose,
Converting rays into sugars sweet,
Carbon's magic, a divine treat.

From the depths of soil it springs,
Roots reaching out, like nature's wings,
Connecting all living, bound by its charm,
Carbon's embrace, a celestial farm.

In diamonds, it lies, a treasure rare,
Sparkling beauty, beyond compare,
Carbon's allure, a mesmerizing sight,
A symbol of love, in the darkest night.

Oh, Carbon, you are versatile indeed,
In every form, a vital need,
From the air we breathe, to the fires that burn,
Carbon's essence, forever we yearn.

In molecules, it weaves, a delicate dance,
Building blocks of life, with every chance,
Proteins, DNA, and all that is alive,
Carbon's touch, a gift we survive.

So let us celebrate this element true,
Carbon, the essence, in all we do,
For in its bonds, life's story unfolds,
A miracle of nature, forever untold.

TWENTY-FIVE

MANY WAYS

Carbon, oh Carbon, so versatile,
In every molecule, you make us smile.
From diamonds to graphite, coal to ink,
You're in everything, it seems, we think.

You bind with oxygen to make CO_2,
A gas that trees need to help them renew.
In proteins and DNA, you play a role,
Without you, life would surely take its toll.

Your allure is undeniable, your presence divine,
In every living thing, you help to define.
So let us celebrate your essence and praise,
Carbon, the element that shapes our fate in so many ways.

TWENTY-SIX

SECRET OF LIFE

In the depths of creation, Carbon abides,
An element versatile, where life resides.
From diamonds to graphite, its forms unfold,
A wealth of wonders, untold and untold.

In the heart of a star, it first takes its birth,
Fusion and fire, a celestial mirth.
Nurtured in cosmic crucibles, it is born,
A building block of life, from dusk till dawn.

In the depths of the Earth, it lies unseen,
Coal and oil, treasures of blackened sheen.
Through time, it transforms, a patient art,
Beneath the ground, it waits to depart.

In the whispers of trees, Carbon finds its voice,
Absorbing the Sun's rays, a choice without choice.
Through photosynthesis, it weaves a song,
Green leaves capturing life, all along.

In the dance of molecules, Carbon takes part,
Bonding and sharing, a molecular art.
Complex chains and structures, it can create,
The architecture of life, intricate and innate.

In the breath we exhale, Carbon takes flight,
A cycle unending, both day and night.
From the depths of the ocean to the heights of the sky,
Carbon connects us all, you and I.

So let us marvel at Carbon's grand design,
A symphony of atoms, interconnected and fine.
For in its essence, lies the secret of life,
Carbon, the element, with beauty rife.

TWENTY-SEVEN

LEARN AND TO GROW

Carbon, oh Carbon, you're a wonder to behold,
A versatile element with stories yet untold.
From diamonds to graphite, you take on many forms,
And in each, you reveal your unique chemical norms.

In nature, you're abundant, in compounds you're diverse,
You're the backbone of life, without you, we'd be cursed.
In proteins and DNA, you play a crucial role,
And in the cycle of life, you help us all stay whole.

But it's not just biology where you shine bright,
In industries and technologies, you're a source of might.
From fuel to plastics to electronics, you're everywhere,
Your presence in our lives, we can hardly compare.

So here's to you, Carbon, a true chemical gem,
A vital part of our world, from beginning to end.
May we continue to study, to learn and to grow,
And uncover all the secrets you have yet to show.

TWENTY-EIGHT

DAZZLING DREAM

In the realm of elements, Carbon reigns supreme,
A versatile atom, like a dazzling dream.
With six protons and electrons, it stands tall,
Forming bonds with ease, connecting one and all.

In diamonds, it sparkles, a crystal so pure,
Reflecting the light, elegance to endure.
Its structure so tight, it captures our gaze,
A symbol of beauty, in infinite ways.

But Carbon's story doesn't end there,
It weaves through compounds with utmost care.
In graphite, it lingers, layer upon layer,
A soft, dark substance, a scribe's favorite player.

In the realm of life, Carbon takes the lead,
As the building block of every living creed.
From plants to animals, all creatures adore,
This element's prowess, forevermore.

In the cycle of life, Carbon plays its part,
From the depths of the oceans to the beating heart.
It cycles through nature, a dance so fine,
Supporting existence, a grand design.
So let us marvel at Carbon's grace,
Its presence in all forms, in every place.
From the depths of the Earth to the skies above,
Carbon, the element we cannot get enough of.

TWENTY-NINE

WITHOUT YOU

In the realm of chemistry, a star does shine,
A building block of life, so divine.
Carbon, the element, so versatile and true,
With bonds that connect, creating something new.

In compounds, it dances, with atoms it entwines,
Forming chains and rings, in a symphony of design.
From diamonds to graphite, it takes many forms,
Carbon, the element, weathers life's storms.

In the depths of the Earth, it lies hidden and deep,
Coal, oil, and gas, where treasures do sleep.
Through heat and pressure, it transforms with grace,
Carbon, the element, holds secrets in its embrace.

In the air we breathe, it is present and free,
Carbon dioxide, a gas that fuels life's decree.
Through photosynthesis, it nourishes the green,
Carbon, the element, in nature's grand scheme.

In biology's realm, it's the backbone of life,
A molecule's scaffold, the essence of strife.
From DNA to proteins, it weaves the code,
Carbon, the element, on which life bestowed.

In industries and technologies, it finds its place,
From plastics to fuels, it leaves its trace.
Carbon, the element, driving innovation's quest,
With endless possibilities, it never rests.

Oh Carbon, the element, so vital and grand,
In every living thing, you take a stand.
From the depths of the Earth to the heights of the sky,
Carbon, the element, without you, life would die.

THIRTY

STAR OF CHEMISTRY'S COSMIC EMBRACE

In the realm of elements, Carbon takes its reign,
A versatile atom, its powers unrestrained.
With four valence electrons, it seeks connection,
Building compounds with grace, a chemical reflection.

From diamonds so pure, sparkling with light,
To graphite's dark hue, a soft, humble sight.
Carbon weaves its magic, in structures so vast,
In every living being, its presence steadfast.

In organic compounds, it takes center stage,
Forming the backbone of life's intricate page.
Carbohydrates, proteins, and lipids galore,
Carbon's bonds hold them together, forevermore.

In carbon dioxide, it rides through the air,
A vital component, in Earth's cycle we share.

Photosynthesis breathes, with Carbon as its fuel,
Green leaves dance with joy, in nature's grand duel.

Petroleum and coal, gifts from ancient decay,
Harnessing Carbon's power, our industries sway.
From plastics to fuels, it shapes our modern age,
Carbon's fingerprints, on every technology stage.

But Carbon's true essence lies in its connection,
To all living creatures, a biological affection.
In DNA's double helix, it plays a crucial role,
Coding life's blueprint, an eternal scroll.

So let us celebrate Carbon, our elemental friend,
A master of creation, on which life depends.
With its versatility, it shapes our world with grace,
Carbon, the star of chemistry's cosmic embrace.

THIRTY-ONE

VERSATILITY

In nature's grand design, a jewel we find,
A mystic element, both strong and kind.
Carbon, the architect of life's grand show,
A dance of atoms, a symphony to bestow.

Within the realms of biology it thrives,
Building blocks of life, where it contrives.
From humble cells to towering trees,
Carbon weaves its magic, with graceful ease.

Industries, they harness its potent might,
Transforming it with fire, heat, and light.
Coal and diamonds, forged in its embrace,
Carbon's versatility, a gift of grace.

From pencils to plastics, it shapes the world,
Threads of carbon, in flags unfurled.
Graphene's promise, a future yet unseen,
Carbon's potential, a constant dream.

So let us marvel at this element so grand,
Carbon, the backbone of life's own band.
From the depths of Earth to the stars above,
Carbon's story, a testament of love.

THIRTY-TWO

ETERNAL ALLURE

In the realm of elements, Carbon shines bright,
A versatile soul, a magnificent sight.
Biology's backbone, the essence of life,
Carbon weaves the web, erasing all strife.

In diamonds it sparkles, with elegance and grace,
Carbon's allure, no one can erase.
From coal to graphite, it changes its form,
A chameleon element, defying the norm.

Industries thrive on Carbon's embrace,
Steel and plastics, it plays a key base.
Technology leaps forward, thanks to its might,
Carbon nanotubes, a futuristic delight.

In DNA's helix, Carbon's presence is found,
Building blocks of life, forever renowned.
Chemical bonds, strong and secure,
Carbon's touch, an eternal allure.

So let us celebrate this element of worth,
Carbon, the foundation of life on Earth.
From the depths of the ocean to the stars above,
Carbon's beauty, an everlasting love.

THIRTY-THREE

DEFIES LIMITATION

In the realm of chemistry, a marvel we find,
A versatile element, of a unique kind.
Carbon, the foundation, the building block,
In a world of compounds, it holds a strong stock.

From diamonds to graphite, its forms so diverse,
Carbon's allure, it only continues to disperse.
In the depths of the Earth, where pressure is high,
It transforms into diamonds, catching the eye.

But carbon's wonders don't end with its shine,
In industries and technologies, it's truly divine.
In steel and alloys, it lends its strength,
Creating structures that go to great lengths.

In the world of plastics, carbon's the key,
Synthetic fibers and polymers, for all to see.
From bottles to fabrics, it shapes our lives,
Carbon's presence, in every stride.

And let's not forget, biology's realm,
Where carbon plays a pivotal helm.
In DNA and proteins, it takes the lead,
A fundamental element, in every living creed.

So let us marvel at carbon's might,
Its versatility, shining so bright.
From nature's embrace to man's innovation,
Carbon, the element that defies limitation.

THIRTY-FOUR

WEAVES ITS MAGIC

In the depths of Earth's embrace, Carbon lies,
A versatile element, a wonder in disguise.
From the stars it descended, a celestial birth,
To shape the world we know, its intrinsic worth.

In industries it thrives, a cornerstone so grand,
Steel and alloys, crafted by its skilled hand.
From skyscrapers towering, to bridges strong and true,
Carbon's strength and resilience, forever shining through.

Plastics and polymers, a world of endless forms,
Carbon's bonds entwined, creating shapes of norms.
From bottles to electronics, it finds its way,
A material of choice, in our modern day.

But beyond the man-made, Carbon's secrets lie,
In the realms of biology, where life does rely.

In every living creature, from the tiniest cell,
Carbon weaves its magic, a tale to tell.
  In plants, it captures sunlight, through photosynthesis,
Creating sustenance, a cycle of life's bliss.
In animals, it forms the building blocks of life,
From bones to muscles, in harmony so rife.
  So let us celebrate this element divine,
Carbon, the foundation, on which we all recline.
From industries to technologies, biology too,
Carbon shapes the world, eternally true.

THIRTY-FIVE

EVER-PRESENT

In nature's realm, Carbon weaves its tale,
A humble element, yet so grand and frail.
From the depths of Earth to the heavens above,
It whispers secrets of life and everlasting love.

In the fiery core, where creation begins,
Carbon dances with atoms, a symphony it spins.
In each living organism, it finds its home,
As the backbone of life, it doesn't roam.

In the diamond's brilliance, Carbon gleams,
A testament to its strength, beyond our dreams.
It sparkles and dazzles, a treasure so rare,
A symbol of endurance, forever to wear.

In industries and technologies, Carbon takes flight,
Graphene and nanotubes, innovation's delight.
It paves the way for progress, with every stride,
A catalyst for change, standing tall with pride.

From fuel and plastics to solar power's might,
Carbon fuels our world, shining bright.
It powers our machines, drives our cars,
A silent hero, reaching for the stars.

Oh, Carbon, you're the magician of elements,
Invisible yet ever-present, with no restraints.
You bind and connect, in a delicate dance,
Bringing harmony and balance, by mere chance.

So let us celebrate, this wondrous Carbon,
For its versatility and strength, we are all fond.
In nature, industries, and technologies, it thrives,
A symbol of life, connecting our lives.

THIRTY-SIX

BACKBONE

Infinite structures, Carbon can form,
A versatile element, in every norm.
From diamonds to graphite, it's all the same,
Carbon's adaptability, never to blame.

In industries and tech, it plays a key role,
Carbon fibers, nanotubes, and more, a whole.
In biology too, it's essential indeed,
From DNA to amino acids, it's a need.

Carbon is a building block, for man-made materials,
From plastics to rubber, it's in all essentials.
Carbon black, a pigment, so dark and deep,
From tires to ink, it's in products we keep.

Carbon fuels progress, innovation, and more,
From coal to oil, it's at the core.
Renewable energy, it's paving the way,
For a brighter future, every day.

Carbon connects, brings harmony to life,
In every aspect, it has no strife.
A humble element, so often overlooked,
Carbon, the backbone, on which life is hooked.

ABOUT THE AUTHOR

Walter the Educator is one of the pseudonyms for Walter Anderson. Formally educated in Chemistry, Business, and Education, he is an educator, an author, a diverse entrepreneur, and the son of a disabled war veteran. "Walter the Educator" shares his time between educating and creating. He holds interests and owns several creative projects that entertain, enlighten, enhance, and educate, hoping to inspire and motivate you.

Follow, find new works, and stay up to date
with Walter the Educator™
at www.WaltertheEducator.com

www.ingramcontent.com/pod-product-compliance
Lightning Source LLC
LaVergne TN
LVHW051959060526
838201LV00059B/3728